# Century of Moon Phases

## *2001 - 2100*

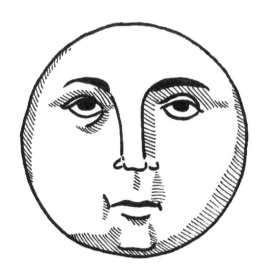

By Sage Liskey
Published by the Rad Cat Press
December 2016

*Illustration: Georgina Tyson*
*Instagram: Cosmicgypotattoo*

First Edition, December, 2016

Published in the United States by the Rad Cat Press.
Written and designed by Sage Liskey.

Printed in the United States of America
1st Edition

ISBN 9780986246135

www.sageliskey.com
www.facebook.com/radcatpress
radcatpress@riseup.net

# About This Book

we wanted to have
a conversation with
Mother Moon and Father Time

*Each image of the moon corresponds to its fullness for Pacific Standard Time.*
*The black moon being the new moon, the white moon being the full moon,*
*and those in between the waxing and waning half moons.*

Front Cover Illustration: Charles Robinson, *A Child's Garden of Verse*
Back Cover Illustration: Anton Steinhauser, *Grundzüge der*
*mathematischen Geographie und der Landkartenprojection*

*Moon Phases Table courtesy of Fred Espenak, www.Astropixels.com*

---

*Illustration: Narrative and Critical History of American*
*(New York: Houghton, Mifflin, and Company, 1886)II:96*

*We are going to the moon that is not very far.*
*Man has so much farther to go within himself.*

—Anaïs Nin

*Illustration: The Half Hour Library of Travel,*
*Nature and Science for Young Readers Vol. 3*

| _The Moon is believed to have formed 4.5 billion years ago._ | | | | |
|---|---|---|---|---|
| | **2001** | | | |
| January | 24 | 2 | 9 | 16 |
| February | 23 | 1 | 8 | 14 |
| March | 24 | 2 | 9 | 16 |
| April | 23 | 1, 30 | 7 | 15 |
| May | 22 | 29 | 7 | 15 |
| June | 21 | 27 | 5 | 13 |
| July | 20 | 27 | 5 | 13 |
| August | 18 | 25 | 3 | 12 |
| September | 17 | 24 | 2 | 10 |
| October | 16 | 23 | 2, 31 | 9 |
| November | 14 | 22 | 30 | 8 |
| December | 14 | 22 | 30 | 7 |
| | **2002** | | | |
| January | 13 | 21 | 28 | 5 |
| February | 12 | 20 | 27 | 4 |
| March | 13 | 21 | 28 | 5 |
| April | 12 | 20 | 26 | 4 |
| May | 12 | 19 | 26 | 4 |
| June | 10 | 17 | 24 | 2 |
| July | 10 | 16 | 24 | 2 |
| August | 8 | 15 | 22 | 1, 30 |
| September | 6 | 13 | 21 | 29 |
| October | 6 | 12 | 21 | 28 |
| November | 4 | 11 | 19 | 27 |
| December | 4 | 11 | 19 | 26 |

| *In Greek mythology, Selene is the goddess of the Moon.* | | | | |
|---|---|---|---|---|
| | 2003 | | | |
| January | 2 | 10 | 18 | 25 |
| February | 1 | 9 | 16 | 23 |
| March | 2 | 11 | 18 | 24 |
| April | 1 | 9 | 16 | 23 |
| May | 1, 30 | 9 | 15 | 22 |
| June | 29 | 7 | 14 | 21 |
| July | 28 | 6 | 13 | 21 |
| August | 27 | 5 | 12 | 19 |
| September | 25 | 3 | 10 | 18 |
| October | 25 | 2, 31 | 10 | 18 |
| November | 23 | 30 | 8 | 16 |
| December | 23 | 30 | 8 | 16 |
| | 2004 | | | |
| January | 21 | 28 | 7 | 14 |
| February | 20 | 27 | 6 | 13 |
| March | 20 | 28 | 6 | 13 |
| April | 19 | 27 | 5 | 11 |
| May | 18 | 27 | 4 | 11 |
| June | 17 | 25 | 2 | 9 |
| July | 17 | 24 | 2, 31 | 9 |
| August | 15 | 23 | 29 | 7 |
| September | 14 | 21 | 28 | 6 |
| October | 13 | 20 | 27 | 6 |
| November | 12 | 18 | 26 | 4 |
| December | 11 | 18 | 26 | 4 |

| The Moon formed when a planetary body known as Theia struck Earth. | | | | |
|---|---|---|---|---|
| | 2005 | | | |
| January | 10 | 16 | 25 | 3 |
| February | 8 | 15 | 23 | 2 |
| March | 10 | 17 | 25 | 3 |
| April | 8 | 16 | 24 | 1, 30 |
| May | 8 | 16 | 23 | 30 |
| June | 6 | 14 | 21 | 28 |
| July | 6 | 14 | 21 | 27 |
| August | 4 | 12 | 19 | 26 |
| September | 3 | 11 | 17 | 24 |
| October | 3 | 10 | 17 | 24 |
| November | 1 | 8 | 15 | 23 |
| December | 1, 30 | 8 | 15 | 23 |
| | 2006 | | | |
| January | 29 | 6 | 14 | 22 |
| February | 27 | 4 | 12 | 21 |
| March | 29 | 6 | 14 | 22 |
| April | 27 | 5 | 13 | 20 |
| May | 26 | 4 | 12 | 20 |
| June | 25 | 3 | 11 | 18 |
| July | 24 | 3 | 10 | 17 |
| August | 23 | 2, 31 | 9 | 15 |
| September | 22 | 30 | 7 | 14 |
| October | 21 | 29 | 6 | 13 |
| November | 20 | 27 | 5 | 12 |
| December | 20 | 27 | 4 | 12 |

| *In Greek mythology, Theia gave birth to Selene.* | | | | |
|---|---|---|---|---|
| | 2007 | | | |
| January | 18 | 25 | 3 | 11 |
| February | 17 | 24 | 1 | 10 |
| March | 18 | 25 | 3 | 11 |
| April | 17 | 23 | 2 | 10 |
| May | 16 | 23 | 2, 31 | 9 |
| June | 14 | 22 | 30 | 8 |
| July | 14 | 21 | 29 | 7 |
| August | 12 | 20 | 28 | 5 |
| September | 11 | 19 | 26 | 3 |
| October | 10 | 19 | 25 | 3 |
| November | 9 | 17 | 24 | 1 |
| December | 9 | 17 | 23 | 1, 31 |
| | 2008 | | | |
| January | 8 | 15 | 22 | 29 |
| February | 6 | 13 | 20 | 28 |
| March | 7 | 14 | 21 | 29 |
| April | 5 | 12 | 20 | 28 |
| May | 5 | 11 | 19 | 27 |
| June | 3 | 10 | 18 | 26 |
| July | 2 | 9 | 18 | 25 |
| August | 1, 30 | 8 | 16 | 23 |
| September | 29 | 7 | 15 | 21 |
| October | 28 | 7 | 14 | 21 |
| November | 27 | 5 | 12 | 19 |
| December | 27 | 5 | 12 | 19 |

| *The collision between Earth and Theia combined parts of both bodies together.* | | | | |
|---|---|---|---|---|
| | 2009 | | | |
| January | 26 | 4 | 10 | 17 |
| February | 24 | 2 | 9 | 16 |
| March | 26 | 4 | 10 | 18 |
| April | 24 | 2 | 9 | 17 |
| May | 24 | 1, 30 | 8 | 17 |
| June | 22 | 29 | 7 | 15 |
| July | 21 | 28 | 7 | 15 |
| August | 20 | 27 | 5 | 13 |
| September | 18 | 25 | 4 | 11 |
| October | 17 | 25 | 3 | 11 |
| November | 16 | 24 | 2 | 9 |
| December | 16 | 24 | 2, 31 | 8 |
| | 2010 | | | |
| January | 15 | 23 | 29 | 7 |
| February | 13 | 21 | 28 | 5 |
| March | 15 | 23 | 29 | 7 |
| April | 14 | 21 | 28 | 6 |
| May | 13 | 20 | 27 | 5 |
| June | 12 | 18 | 26 | 4 |
| July | 11 | 18 | 25 | 4 |
| August | 9 | 16 | 24 | 2 |
| September | 8 | 14 | 23 | 1, 30 |
| October | 7 | 14 | 22 | 30 |
| November | 5 | 13 | 21 | 28 |
| December | 5 | 13 | 21 | 27 |

*Illustration: A.B. Woodward, Red Apple and Silver Bells*

| *The compositions of the Moon and Earth are similar.* | | | | |
|---|---|---|---|---|
| | 2011 | | | |
| January | 4 | 12 | 19 | 26 |
| February | 2 | 11 | 18 | 24 |
| March | 4 | 12 | 19 | 26 |
| April | 3 | 11 | 17 | 24 |
| May | 2 | 10 | 17 | 24 |
| June | 1 | 8 | 15 | 23 |
| July | 1, 30 | 7 | 14 | 22 |
| August | 28 | 6 | 13 | 21 |
| September | 27 | 4 | 12 | 20 |
| October | 26 | 3 | 11 | 19 |
| November | 24 | 2 | 10 | 18 |
| December | 24 | 2, 31 | 10 | 17 |
| | 2012 | | | |
| January | 23 | 30 | 9 | 16 |
| February | 21 | 29 | 7 | 14 |
| March | 22 | 30 | 8 | 14 |
| April | 21 | 29 | 6 | 13 |
| May | 20 | 28 | 5 | 12 |
| June | 19 | 26 | 4 | 11 |
| July | 18 | 26 | 3 | 10 |
| August | 17 | 24 | 1, 31 | 9 |
| September | 15 | 22 | 29 | 8 |
| October | 15 | 21 | 29 | 8 |
| November | 13 | 20 | 28 | 6 |
| December | 13 | 19 | 28 | 6 |

| *43% oxygen, 20% silicon, 19% magnesium, 10% iron, 3% calcium, 3% aluminum.* | | | |
|---|---|---|---|
| **2013** | | | |
| January | 11 | 18 | 26 | 4 |
| February | 10 | 17 | 25 | 3 |
| March | 11 | 19 | 27 | 4 |
| April | 10 | 18 | 25 | 2 |
| May | 9 | 17 | 24 | 2, 31 |
| June | 8 | 16 | 23 | 29 |
| July | 8 | 15 | 22 | 29 |
| August | 6 | 14 | 20 | 28 |
| September | 5 | 12 | 19 | 26 |
| October | 4 | 11 | 18 | 26 |
| November | 3 | 9 | 17 | 25 |
| December | 2 | 9 | 17 | 25 |
| **2014** | | | |
| January | 1, 30 | 7 | 15 | 23 |
| February | | 6 | 14 | 22 |
| March | 1, 30 | 8 | 16 | 23 |
| April | 28 | 7 | 15 | 22 |
| May | 28 | 6 | 14 | 21 |
| June | 27 | 5 | 12 | 19 |
| July | 26 | 5 | 12 | 18 |
| August | 25 | 3 | 10 | 17 |
| September | 23 | 2 | 8 | 15 |
| October | 23 | 1, 30 | 8 | 15 |
| November | 22 | 29 | 6 | 14 |
| December | 21 | 28 | 6 | 14 |

| 2,159.2 miles (3,475 kilometers) in diameter. | | | |
|---|---|---|---|
| 2015 | | | |
| January | 20 | 26 | 4 | 13 |
| February | 18 | 25 | 3 | 11 |
| March | 20 | 27 | 5 | 13 |
| April | 18 | 25 | 4 | 11 |
| May | 17 | 25 | 3 | 11 |
| June | 16 | 24 | 2 | 9 |
| July | 15 | 23 | 1, 31 | 8 |
| August | 14 | 22 | 29 | 6 |
| September | 12 | 21 | 27 | 5 |
| October | 12 | 20 | 27 | 4 |
| November | 11 | 18 | 25 | 3 |
| December | 11 | 18 | 25 | 3 |
| 2016 | | | |
| January | 9 | 16 | 23 | 1, 31 |
| February | 8 | 15 | 22 | |
| March | 8 | 15 | 23 | 1, 31 |
| April | 7 | 13 | 21 | 29 |
| May | 6 | 13 | 21 | 29 |
| June | 4 | 12 | 20 | 27 |
| July | 4 | 11 | 19 | 26 |
| August | 2 | 10 | 18 | 24 |
| September | 1, 30 | 9 | 16 | 23 |
| October | 30 | 8 | 15 | 22 |
| November | 29 | 7 | 14 | 21 |
| December | 28 | 7 | 13 | 20 |

| 6,783.5 miles (10,917 kilometers) in equatorial circumference. | | | | |
|---|---|---|---|---|
| | | **2017** | | |
| January | 27 | 5 | 12 | 19 |
| February | 26 | 3 | 10 | 18 |
| March | 27 | 5 | 12 | 20 |
| April | 26 | 3 | 10 | 19 |
| May | 25 | 2 | 10 | 18 |
| June | 23 | 1, 30 | 9 | 17 |
| July | 23 | 30 | 8 | 16 |
| August | 21 | 29 | 7 | 14 |
| September | 19 | 27 | 6 | 12 |
| October | 19 | 27 | 5 | 12 |
| November | 18 | 26 | 3 | 10 |
| December | 17 | 26 | 3 | 10 |
| | | **2018** | | |
| January | 16 | 24 | 1, 31 | 8 |
| February | 15 | 23 | | 7 |
| March | 17 | 24 | 1, 31 | 9 |
| April | 15 | 22 | 29 | 8 |
| May | 15 | 21 | 29 | 7 |
| June | 13 | 20 | 27 | 6 |
| July | 12 | 19 | 27 | 6 |
| August | 11 | 18 | 26 | 4 |
| September | 9 | 16 | 24 | 2 |
| October | 8 | 16 | 24 | 2, 31 |
| November | 7 | 15 | 22 | 29 |
| December | 7 | 15 | 22 | 29 |

| It takes 27 days, 7 hours, and 43.1 minutes to orbit around the Earth. | | | | |
|---|---|---|---|---|
| | 2019 | | | |
| January | 5 | 13 | 20 | 27 |
| February | 4 | 12 | 19 | 26 |
| March | 6 | 14 | 20 | 27 |
| April | 5 | 12 | 19 | 26 |
| May | 4 | 11 | 18 | 26 |
| June | 3 | 9 | 17 | 25 |
| July | 2, 31 | 9 | 16 | 24 |
| August | 30 | 7 | 15 | 23 |
| September | 28 | 5 | 13 | 21 |
| October | 27 | 5 | 13 | 21 |
| November | 26 | 4 | 12 | 19 |
| December | 25 | 3 | 11 | 18 |
| | 2020 | | | |
| January | 24 | 2 | 10 | 17 |
| February | 23 | 1 | 9 | 15 |
| March | 24 | 2 | 9 | 16 |
| April | 22 | 1, 30 | 7 | 14 |
| May | 22 | 29 | 7 | 14 |
| June | 20 | 28 | 5 | 12 |
| July | 20 | 27 | 4 | 12 |
| August | 18 | 25 | 3 | 11 |
| September | 17 | 23 | 1 | 10 |
| October | 16 | 23 | 1, 31 | 9 |
| November | 14 | 21 | 30 | 8 |
| December | 14 | 21 | 29 | 7 |

We are all like the bright moon,
we still have our darker side.

—Kahlil Gibran

Illustration: Klugh, Maria Tales from the Far North
(Chicago, IL: A. Flanagan Company, 1909)

| *Due to the spin of Earth, the Moon takes 29.5 days to fully rotate for Earthlings.* | | | | |
|---|---|---|---|---|
| | **2021** | | | |
| January | 12 | 20 | 28 | 6 |
| February | 11 | 19 | 27 | 4 |
| March | 13 | 21 | 28 | 5 |
| April | 11 | 19 | 26 | 4 |
| May | 11 | 19 | 26 | 3 |
| June | 10 | 17 | 24 | 2 |
| July | 9 | 17 | 23 | 1, 31 |
| August | 8 | 15 | 22 | 30 |
| September | 6 | 13 | 20 | 28 |
| October | 6 | 12 | 20 | 28 |
| November | 4 | 11 | 19 | 27 |
| December | 4 | 10 | 18 | 26 |
| | **2022** | | | |
| January | 2, 31 | 9 | 17 | 25 |
| February | | 8 | 16 | 23 |
| March | 2, 31 | 10 | 18 | 24 |
| April | 30 | 8 | 16 | 23 |
| May | 30 | 8 | 15 | 22 |
| June | 28 | 7 | 14 | 20 |
| July | 28 | 6 | 13 | 20 |
| August | 27 | 5 | 11 | 18 |
| September | 25 | 3 | 10 | 17 |
| October | 25 | 2, 31 | 9 | 17 |
| November | 23 | 30 | 8 | 16 |
| December | 23 | 29 | 7 | 16 |

| *Moon was originally derived from the Proto-Germanic *mǣnōn.* | | | | |
|---|---|---|---|---|
| | | 2023 | | |
| January | 21 | 28 | 6 | 14 |
| February | 20 | 27 | 5 | 13 |
| March | 21 | 28 | 7 | 14 |
| April | 19 | 27 | 5 | 13 |
| May | 19 | 27 | 5 | 12 |
| June | 17 | 26 | 3 | 10 |
| July | 17 | 25 | 3 | 9 |
| August | 16 | 24 | 1, 30 | 8 |
| September | 14 | 22 | 29 | 6 |
| October | 14 | 21 | 28 | 6 |
| November | 13 | 20 | 27 | 5 |
| December | 12 | 19 | 26 | 4 |
| | | 2024 | | |
| January | 11 | 17 | 25 | 3 |
| February | 9 | 16 | 24 | 2 |
| March | 10 | 16 | 25 | 3 |
| April | 8 | 15 | 23 | 1 |
| May | 7 | 15 | 23 | 1, 30 |
| June | 6 | 13 | 21 | 28 |
| July | 5 | 13 | 21 | 27 |
| August | 4 | 12 | 19 | 26 |
| September | 2 | 10 | 17 | 24 |
| October | 2 | 10 | 17 | 24 |
| November | 1, 30 | 8 | 15 | 22 |
| December | 30 | 8 | 15 | 22 |

| *After *mǣnōn., the spelling changed to mōna, mone, moone, then moon.* | | | | |
|---|---|---|---|---|
| | **2025** | | | |
| January | 29 | 6 | 13 | 21 |
| February | 27 | 5 | 12 | 20 |
| March | 29 | 6 | 13 | 22 |
| April | 27 | 4 | 12 | 20 |
| May | 26 | 4 | 12 | 20 |
| June | 25 | 2 | 11 | 18 |
| July | 24 | 2 | 10 | 17 |
| August | 22 | 1, 31 | 9 | 15 |
| September | 21 | 29 | 7 | 14 |
| October | 21 | 29 | 6 | 13 |
| November | 19 | 27 | 5 | 11 |
| December | 19 | 27 | 4 | 11 |
| | **2026** | | | |
| January | 18 | 25 | 3 | 10 |
| February | 17 | 24 | 1 | 9 |
| March | 18 | 25 | 3 | 11 |
| April | 17 | 23 | 1 | 9 |
| May | 16 | 23 | 1, 31 | 9 |
| June | 14 | 21 | 29 | 8 |
| July | 14 | 21 | 29 | 7 |
| August | 12 | 19 | 27 | 5 |
| September | 10 | 18 | 26 | 4 |
| October | 10 | 18 | 25 | 3 |
| November | 9 | 17 | 24 | 1, 30 |
| December | 8 | 16 | 23 | 30 |

| *5400 square miles of the surface is in permanent darkness.* | | | | |
|---|---|---|---|---|
| | **2027** | | | |
| January | 7 | 15 | 22 | 29 |
| February | 6 | 14 | 20 | 27 |
| March | 8 | 15 | 22 | 29 |
| April | 6 | 13 | 20 | 28 |
| May | 6 | 12 | 20 | 28 |
| June | 4 | 11 | 18 | 26 |
| July | 3 | 10 | 18 | 26 |
| August | 2, 31 | 8 | 17 | 24 |
| September | 29 | 7 | 15 | 23 |
| October | 29 | 7 | 15 | 22 |
| November | 27 | 6 | 13 | 20 |
| December | 27 | 5 | 13 | 20 |
| | **2028** | | | |
| January | 26 | 4 | 11 | 18 |
| February | 25 | 3 | 10 | 17 |
| March | 25 | 4 | 10 | 17 |
| April | 24 | 2 | 9 | 16 |
| May | 24 | 1, 31 | 8 | 16 |
| June | 22 | 29 | 6 | 14 |
| July | 21 | 28 | 6 | 14 |
| August | 20 | 26 | 5 | 13 |
| September | 18 | 25 | 3 | 11 |
| October | 17 | 24 | 3 | 11 |
| November | 16 | 23 | 2 | 9 |
| December | 15 | 23 | 1, 31 | 8 |

| *In 1959 the Soviet Union's Luna 2 was the first robot ever landed on the Moon's surface.* | | | | |
|---|---|---|---|---|
| | | 2029 | | |
| January | 14 | 22 | 29 | 7 |
| February | 13 | 21 | 28 | 5 |
| March | 14 | 23 | 29 | 7 |
| April | 13 | 21 | 28 | 5 |
| May | 13 | 20 | 27 | 5 |
| June | 11 | 19 | 25 | 3 |
| July | 11 | 18 | 25 | 3 |
| August | 9 | 16 | 23 | 2, 31 |
| September | 8 | 14 | 22 | 30 |
| October | 7 | 14 | 22 | 30 |
| November | 5 | 12 | 20 | 28 |
| December | 5 | 12 | 20 | 28 |
| | | 2030 | | |
| January | 3 | 11 | 19 | 26 |
| February | 2 | 10 | 17 | 24 |
| March | 3 | 12 | 19 | 26 |
| April | 2 | 10 | 17 | 24 |
| May | 2, 31 | 10 | 17 | 23 |
| June | 30 | 8 | 15 | 22 |
| July | 30 | 8 | 14 | 22 |
| August | 28 | 6 | 13 | 20 |
| September | 27 | 4 | 11 | 19 |
| October | 26 | 3 | 11 | 19 |
| November | 24 | 2 | 9 | 18 |
| December | 24 | 1, 31 | 9 | 17 |

*The moon does not fight. It attacks no one. It does not worry. It does not try to crush others. It keeps to its course, but by its very nature, it gently influences. What other body could pull an entire ocean from shore to shore? The moon is faithful to its nature and its power is never diminished.*

—Ming-Dao Deng, *Everyday Tao: Living with Balance and Harmony*

*Illustration: H.J. Ford and Lancelot Speed, The Blue Poetry Book*

| *In 1968 the USA launched the Apollo 8, the first manned moon mission.* | | | | |
|---|---|---|---|---|
| | 2031 | | | |
| January | 22 | 30 | 8 | 16 |
| February | 21 | 28 | 7 | 14 |
| March | 22 | 30 | 8 | 15 |
| April | 21 | 29 | 7 | 14 |
| May | 21 | 29 | 6 | 13 |
| June | 19 | 27 | 5 | 11 |
| July | 19 | 27 | 4 | 11 |
| August | 17 | 25 | 2 | 9 |
| September | 16 | 23 | 1, 30 | 8 |
| October | 16 | 23 | 30 | 8 |
| November | 14 | 21 | 28 | 7 |
| December | 14 | 20 | 28 | 6 |
| | 2032 | | | |
| January | 12 | 19 | 27 | 5 |
| February | 10 | 17 | 26 | 4 |
| March | 11 | 18 | 26 | 4 |
| April | 9 | 17 | 25 | 3 |
| May | 9 | 17 | 24 | 2, 31 |
| June | 7 | 15 | 23 | 29 |
| July | 7 | 15 | 22 | 29 |
| August | 5 | 14 | 20 | 27 |
| September | 4 | 12 | 19 | 26 |
| October | 4 | 11 | 18 | 25 |
| November | 2 | 10 | 16 | 24 |
| December | 2 | 9 | 16 | 24 |

| *In 1969 Neil Armstrong was the first human to ever walk on the moon.* | | | | |
|---|---|---|---|---|
| | | 2033 | | |
| January | 1, 30 | 7 | 15 | 23 |
| February | | 6 | 14 | 22 |
| March | 1, 30 | 7 | 15 | 23 |
| April | 28 | 6 | 14 | 22 |
| May | 28 | 5 | 14 | 21 |
| June | 26 | 4 | 12 | 19 |
| July | 26 | 4 | 12 | 18 |
| August | 24 | 3 | 10 | 17 |
| September | 23 | 1 | 8 | 15 |
| October | 23 | 1, 30 | 8 | 14 |
| November | 21 | 29 | 6 | 13 |
| December | 21 | 28 | 6 | 13 |
| | | 2034 | | |
| January | 20 | 27 | 4 | 12 |
| February | 18 | 25 | 3 | 11 |
| March | 20 | 26 | 4 | 12 |
| April | 18 | 25 | 3 | 11 |
| May | 17 | 24 | 3 | 11 |
| June | 16 | 23 | 1 | 9 |
| July | 15 | 23 | 1, 30 | 8 |
| August | 13 | 21 | 29 | 6 |
| September | 12 | 20 | 27 | 5 |
| October | 12 | 20 | 27 | 4 |
| November | 10 | 18 | 25 | 2 |
| December | 10 | 18 | 25 | 2 |

| *500 million people watched the live broadcast of the Apollo 8 space landing.* | | | | |
|---|---|---|---|---|
| | | | 2035 | |
| January | 9 | 16 | 23 | 1, 30 |
| February | 8 | 15 | 22 | |
| March | 9 | 16 | 23 | 1, 31 |
| April | 8 | 14 | 22 | 30 |
| May | 7 | 14 | 21 | 30 |
| June | 5 | 12 | 20 | 28 |
| July | 5 | 12 | 20 | 27 |
| August | 3 | 10 | 18 | 26 |
| September | 1 | 9 | 17 | 24 |
| October | 1, 30 | 9 | 16 | 23 |
| November | 29 | 7 | 15 | 21 |
| December | 29 | 7 | 14 | 21 |
| | | | 2036 | |
| January | 28 | 6 | 13 | 19 |
| February | 26 | 5 | 11 | 18 |
| March | 27 | 5 | 12 | 19 |
| April | 26 | 3 | 10 | 18 |
| May | 25 | 2 | 10 | 18 |
| June | 23 | 1, 30 | 8 | 16 |
| July | 23 | 29 | 8 | 16 |
| August | 21 | 28 | 6 | 14 |
| September | 19 | 26 | 5 | 13 |
| October | 19 | 26 | 5 | 12 |
| November | 17 | 25 | 3 | 10 |
| December | 17 | 25 | 3 | 10 |

| There is frozen water on the moon. | | | | |
|---|---|---|---|---|
| **2037** | | | | |
| January | 16 | 24 | 1, 31 | 8 |
| February | 14 | 22 | | 6 |
| March | 16 | 24 | 1, 31 | 8 |
| April | 15 | 22 | 29 | 7 |
| May | 14 | 22 | 28 | 6 |
| June | 13 | 20 | 27 | 5 |
| July | 12 | 19 | 26 | 5 |
| August | 11 | 17 | 25 | 4 |
| September | 9 | 16 | 24 | 2 |
| October | 8 | 15 | 23 | 2, 31 |
| November | 7 | 14 | 22 | 29 |
| December | 6 | 14 | 22 | 29 |
| **2038** | | | | |
| January | 5 | 13 | 20 | 27 |
| February | 3 | 12 | 19 | 25 |
| March | 5 | 13 | 20 | 27 |
| April | 4 | 12 | 19 | 25 |
| May | 4 | 11 | 18 | 25 |
| June | 2 | 10 | 16 | 24 |
| July | 2, 31 | 9 | 16 | 23 |
| August | 30 | 7 | 14 | 22 |
| September | 28 | 5 | 13 | 21 |
| October | 27 | 5 | 12 | 21 |
| November | 26 | 3 | 11 | 19 |
| December | 25 | 3 | 11 | 19 |

| *An eclipse occurs when the Earth, Sun, and Moon all form a straight line.* | | | | |
|---|---|---|---|---|
| | | 2039 | | |
| January | 24 | 2, 31 | 10 | 17 |
| February | 22 | | 8 | 15 |
| March | 24 | 2 | 10 | 17 |
| April | 23 | 1 | 8 | 15 |
| May | 22 | 1, 30 | 8 | 14 |
| June | 21 | 29 | 6 | 13 |
| July | 21 | 28 | 5 | 12 |
| August | 19 | 26 | 4 | 11 |
| September | 18 | 24 | 2 | 10 |
| October | 17 | 24 | 2, 31 | 10 |
| November | 15 | 22 | 30 | 8 |
| December | 15 | 22 | 30 | 8 |
| | | 2040 | | |
| January | 13 | 20 | 29 | 7 |
| February | 12 | 19 | 27 | 5 |
| March | 12 | 20 | 28 | 6 |
| April | 11 | 19 | 26 | 4 |
| May | 10 | 19 | 26 | 3 |
| June | 9 | 17 | 24 | 1 |
| July | 9 | 17 | 23 | 1, 30 |
| August | 7 | 15 | 22 | 29 |
| September | 6 | 13 | 20 | 27 |
| October | 5 | 13 | 19 | 27 |
| November | 4 | 11 | 18 | 26 |
| December | 4 | 10 | 18 | 26 |

The moon is a loyal companion. It never leaves. It's always there, watching, steadfast, knowing us in our light and dark moments, changing forever just as we do. Every day it's a different version of itself. Sometimes weak and wan, sometimes strong and full of light. The moon understands what it means to be human. Uncertain. Alone. Cratered by imperfections.

—*Tahereh Mafi, Shatter Me*

*Illustration: O.J. Wedlandt, Living Statuary*

| Solar eclipses occur when the Moon is between the Sun and Earth. | | | | |
|---|---|---|---|---|
| | 2041 | | | |
| January | 2, 31 | 9 | 17 | 25 |
| February | | 7 | 15 | 23 |
| March | 2, 31 | 9 | 17 | 25 |
| April | 30 | 8 | 16 | 23 |
| May | 29 | 7 | 15 | 22 |
| June | 28 | 6 | 14 | 20 |
| July | 27 | 6 | 13 | 20 |
| August | 26 | 4 | 11 | 18 |
| September | 25 | 3 | 10 | 16 |
| October | 24 | 2 | 9 | 16 |
| November | 23 | 1, 30 | 7 | 15 |
| December | 23 | 29 | 7 | 15 |
| | 2042 | | | |
| January | 21 | 28 | 6 | 14 |
| February | 20 | 26 | 4 | 13 |
| March | 21 | 28 | 6 | 14 |
| April | 19 | 26 | 5 | 13 |
| May | 19 | 26 | 4 | 12 |
| June | 17 | 25 | 3 | 10 |
| July | 16 | 24 | 3 | 9 |
| August | 15 | 23 | 1, 30 | 8 |
| September | 14 | 22 | 29 | 6 |
| October | 13 | 21 | 28 | 5 |
| November | 12 | 20 | 26 | 4 |
| December | 12 | 19 | 26 | 4 |

| *Lunar eclipses occur when the Earth is between the Sun and Moon.* | | | | |
|---|---|---|---|---|
| | **2043** | | | |
| January | 10 | 18 | 24 | 2 |
| February | 9 | 16 | 23 | 1 |
| March | 11 | 17 | 25 | 3 |
| April | 9 | 16 | 24 | 2 |
| May | 8 | 15 | 23 | 2, 31 |
| June | 7 | 14 | 22 | 29 |
| July | 6 | 13 | 21 | 29 |
| August | 4 | 12 | 20 | 27 |
| September | 3 | 11 | 18 | 25 |
| October | 2 | 11 | 18 | 24 |
| November | 1 | 9 | 16 | 23 |
| December | 1, 31 | 9 | 16 | 22 |
| | **2044** | | | |
| January | 29 | 7 | 14 | 21 |
| February | 28 | 6 | 12 | 20 |
| March | 29 | 6 | 13 | 21 |
| April | 27 | 4 | 12 | 20 |
| May | 26 | 4 | 11 | 19 |
| June | 25 | 2 | 10 | 18 |
| July | 24 | 1, 31 | 9 | 17 |
| August | 22 | 30 | 8 | 16 |
| September | 21 | 28 | 7 | 14 |
| October | 20 | 28 | 6 | 13 |
| November | 19 | 27 | 5 | 11 |
| December | 19 | 27 | 4 | 11 |

| *Both types of eclipses occur approximately every 18 years.* | | | | |
|---|---|---|---|---|
| | | | 2045 | |
| January | 17 | 25 | 3 | 9 |
| February | 16 | 24 | 1 | 8 |
| March | 18 | 25 | 3 | 10 |
| April | 17 | 24 | 1, 30 | 9 |
| May | 16 | 23 | 30 | 8 |
| June | 14 | 21 | 29 | 7 |
| July | 14 | 20 | 28 | 7 |
| August | 12 | 19 | 27 | 5 |
| September | 10 | 17 | 25 | 4 |
| October | 10 | 17 | 25 | 3 |
| November | 8 | 16 | 24 | 1 |
| December | 8 | 16 | 23 | 1, 30 |
| | | | 2046 | |
| January | 6 | 15 | 22 | 28 |
| February | 5 | 13 | 20 | 27 |
| March | 7 | 15 | 22 | 28 |
| April | 6 | 13 | 20 | 27 |
| May | 5 | 13 | 19 | 27 |
| June | 4 | 11 | 18 | 26 |
| July | 3 | 10 | 17 | 25 |
| August | 2, 31 | 8 | 16 | 24 |
| September | 29 | 7 | 14 | 23 |
| October | 29 | 6 | 14 | 22 |
| November | 27 | 5 | 13 | 20 |
| December | 27 | 5 | 13 | 20 |

| The 1967 Outer Space Treaty deems space as the province of all mankind. | | | | |
|---|---|---|---|---|
| | | | 2047 | |
| January | 25 | 3 | 11 | 18 |
| February | 24 | 2 | 10 | 16 |
| March | 26 | 4 | 11 | 18 |
| April | 24 | 3 | 10 | 16 |
| May | 24 | 2 | 9 | 16 |
| June | 23 | 1, 30 | 7 | 15 |
| July | 22 | 29 | 7 | 14 |
| August | 21 | 27 | 5 | 13 |
| September | 19 | 26 | 4 | 12 |
| October | 18 | 25 | 3 | 11 |
| November | 17 | 24 | 2 | 10 |
| December | 16 | 23 | 2, 31 | 10 |
| | | | 2048 | |
| January | 15 | 22 | 30 | 8 |
| February | 13 | 21 | 29 | 6 |
| March | 14 | 22 | 29 | 7 |
| April | 12 | 21 | 28 | 5 |
| May | 12 | 20 | 27 | 4 |
| June | 11 | 19 | 25 | 3 |
| July | 10 | 18 | 25 | 2 |
| August | 9 | 16 | 23 | 1, 31 |
| September | 7 | 14 | 21 | 29 |
| October | 7 | 14 | 21 | 29 |
| November | 5 | 12 | 20 | 28 |
| December | 5 | 12 | 19 | 28 |

| *The 1967 Outer Space Treaty prohibits military usage of the Moon.* | | | |
|---|---|---|---|
| **2049** | | | |
| January | 3 | 10 | 18 | 26 |
| February | 2 | 9 | 17 | 25 |
| March | 3 | 11 | 19 | 26 |
| April | 2 | 10 | 17 | 24 |
| May | 1, 31 | 9 | 17 | 23 |
| June | 29 | 8 | 15 | 22 |
| July | 29 | 8 | 14 | 21 |
| August | 28 | 6 | 13 | 20 |
| September | 26 | 4 | 11 | 18 |
| October | 26 | 4 | 10 | 18 |
| November | 24 | 2 | 9 | 17 |
| December | 24 | 1, 31 | 9 | 17 |
| **2050** | | | |
| January | 22 | 29 | 7 | 15 |
| February | 21 | 28 | 6 | 14 |
| March | 22 | 29 | 8 | 16 |
| April | 21 | 28 | 7 | 14 |
| May | 20 | 28 | 6 | 13 |
| June | 19 | 27 | 5 | 11 |
| July | 18 | 26 | 4 | 11 |
| August | 17 | 25 | 2 | 9 |
| September | 15 | 23 | 1, 30 | 7 |
| October | 15 | 23 | 29 | 7 |
| November | 14 | 21 | 28 | 6 |
| December | 13 | 20 | 27 | 5 |

*But even when the moon looks like it's waning...*
*it's actually never changing shape. Don't ever forget that.*

—Ai Yazawa, *Nana, Vol. 14*

| *Many peoples once relied on moon calenders to tell the time of year.* | | | | |
|---|---|---|---|---|
| | 2051 | | | |
| January | 12 | 19 | 26 | 4 |
| February | 10 | 17 | 25 | 3 |
| March | 12 | 19 | 27 | 5 |
| April | 10 | 17 | 25 | 4 |
| May | 10 | 17 | 25 | 3 |
| June | 8 | 15 | 23 | 1 |
| July | 7 | 15 | 23 | 1, 30 |
| August | 6 | 14 | 21 | 28 |
| September | 4 | 13 | 20 | 26 |
| October | 4 | 12 | 19 | 26 |
| November | 3 | 11 | 17 | 24 |
| December | 3 | 10 | 17 | 24 |
| | 2052 | | | |
| January | 1, 31 | 9 | 15 | 23 |
| February | | 7 | 14 | 22 |
| March | 1, 30 | 7 | 15 | 23 |
| April | 28 | 6 | 13 | 21 |
| May | 28 | 5 | 13 | 21 |
| June | 26 | 3 | 12 | 19 |
| July | 25 | 3 | 11 | 19 |
| August | 24 | 2 | 10 | 17 |
| September | 22 | 1, 30 | 8 | 15 |
| October | 22 | 30 | 8 | 14 |
| November | 21 | 29 | 6 | 13 |
| December | 20 | 28 | 6 | 12 |

| The Moon's gravitational pull helps influence ocean waves and tides. | | | | |
|---|---|---|---|---|
| | | | 2053 | |
| January | 19 | 27 | 4 | 11 |
| February | 18 | 25 | 2 | 10 |
| March | 20 | 26 | 4 | 12 |
| April | 18 | 25 | 2 | 10 |
| May | 17 | 24 | 2 | 10 |
| June | 16 | 22 | 1, 30 | 9 |
| July | 15 | 22 | 30 | 8 |
| August | 13 | 20 | 29 | 7 |
| September | 12 | 19 | 27 | 5 |
| October | 11 | 19 | 27 | 4 |
| November | 10 | 18 | 25 | 2 |
| December | 9 | 18 | 25 | 2, 31 |
| | | | 2054 | |
| January | 8 | 16 | 23 | 30 |
| February | 7 | 15 | 21 | |
| March | 9 | 16 | 23 | 1, 30 |
| April | 7 | 15 | 21 | 29 |
| May | 7 | 14 | 21 | 29 |
| June | 5 | 12 | 19 | 28 |
| July | 5 | 11 | 19 | 27 |
| August | 3 | 10 | 18 | 26 |
| September | 1 | 8 | 16 | 24 |
| October | 1, 30 | 8 | 16 | 23 |
| November | 29 | 7 | 15 | 22 |
| December | 28 | 6 | 14 | 21 |

| *Some craters reach temperatures as low as −413 °F (−247 °C).* | | | |
|---|---|---|---|
| | | 2055 | |
| January | 27 | 5 | 13 | 19 |
| February | 26 | 4 | 11 | 18 |
| March | 28 | 6 | 13 | 19 |
| April | 26 | 4 | 11 | 18 |
| May | 26 | 4 | 10 | 18 |
| June | 24 | 2 | 9 | 16 |
| July | 24 | 1, 30 | 8 | 16 |
| August | 22 | 29 | 7 | 15 |
| September | 20 | 27 | 5 | 13 |
| October | 20 | 27 | 5 | 13 |
| November | 18 | 25 | 4 | 11 |
| December | 18 | 25 | 3 | 11 |
| | | 2056 | |
| January | 16 | 24 | 2 | 9 |
| February | 15 | 23 | 1 | 8 |
| March | 15 | 24 | 1, 31 | 8 |
| April | 14 | 22 | 29 | 6 |
| May | 14 | 22 | 28 | 6 |
| June | 13 | 20 | 27 | 4 |
| July | 12 | 19 | 26 | 4 |
| August | 11 | 17 | 24 | 3 |
| September | 9 | 16 | 23 | 1 |
| October | 8 | 15 | 23 | 1, 31 |
| November | 7 | 13 | 21 | 29 |
| December | 6 | 13 | 21 | 29 |

| Parts of the Moon become as hot as 253 °F (123 °C). | | | |
|---|---|---|---|
| | | 2057 | |
| January | 5 | 12 | 20 | 27 |
| February | 3 | 11 | 19 | 26 |
| March | 5 | 13 | 20 | 27 |
| April | 3 | 11 | 19 | 25 |
| May | 3 | 11 | 18 | 25 |
| June | 2 | 10 | 16 | 23 |
| July | 1, 31 | 9 | 16 | 23 |
| August | 29 | 7 | 14 | 21 |
| September | 28 | 6 | 12 | 20 |
| October | 27 | 5 | 12 | 20 |
| November | 26 | 3 | 10 | 19 |
| December | 25 | 2 | 10 | 18 |
| | | 2058 | |
| January | 24 | 1, 31 | 9 | 17 |
| February | 22 | | 8 | 16 |
| March | 24 | 1, 31 | 10 | 17 |
| April | 22 | 30 | 8 | 15 |
| May | 22 | 30 | 8 | 14 |
| June | 20 | 28 | 6 | 13 |
| July | 20 | 28 | 5 | 12 |
| August | 19 | 26 | 4 | 10 |
| September | 17 | 25 | 2 | 9 |
| October | 17 | 24 | 1, 31 | 9 |
| November | 15 | 22 | 29 | 8 |
| December | 15 | 22 | 29 | 7 |

36

| *The average distance to the Earth is 238,900 miles (384,400 kilometers).* | | | | |
|---|---|---|---|---|
| | | **2059** | | |
| January | 13 | 20 | 28 | 6 |
| February | 12 | 19 | 27 | 5 |
| March | 13 | 20 | 28 | 7 |
| April | 12 | 19 | 27 | 5 |
| May | 11 | 19 | 27 | 4 |
| June | 9 | 17 | 25 | 2 |
| July | 9 | 17 | 24 | 2, 31 |
| August | 8 | 16 | 23 | 29 |
| September | 6 | 14 | 21 | 28 |
| October | 6 | 14 | 20 | 27 |
| November | 5 | 12 | 19 | 26 |
| December | 4 | 11 | 18 | 26 |
| | | **2060** | | |
| January | 3 | 10 | 17 | 25 |
| February | 1 | 8 | 16 | 24 |
| March | 2, 31 | 9 | 16 | 25 |
| April | 30 | 7 | 15 | 23 |
| May | 29 | 7 | 15 | 22 |
| June | 27 | 5 | 13 | 21 |
| July | 27 | 5 | 13 | 20 |
| August | 25 | 4 | 11 | 18 |
| September | 24 | 2 | 10 | 16 |
| October | 24 | 2 | 9 | 16 |
| November | 22 | 1, 30 | 7 | 14 |
| December | 22 | 30 | 7 | 14 |

*The moon had been observing the earth close-up longer than anyone. It must have witnessed all of the phenomena occurring and all of the acts carried out on this earth. But the moon remained silent; it told no stories. All it did was embrace the heavy past with a cool, measured detachment. On the moon there was neither air nor wind. Its vacuum was perfect for preserving memories unscathed. No one could unlock the heart of the moon.*

*Aomame raised her glass to the moon and asked,*
*"Have you gone to bed with someone in your arms lately?"*
*The moon did not answer.*
*"Do you have any friends?" she asked.*
*The moon did not answer.*
*"Don't you get tired of always playing it cool?"*
*The moon did not answer.*

—Haruki Murakami, *1Q84*

---

*Illustration: The Half Hour Library of Travel,*
*Nature and Science for Young Readers Vol. 3*

| *Every year the Moon drifts 1.5 inches (3.8 centimeters) further away from the Earth.* | | | |
|---|---|---|---|
| | | 2061 | |
| January | 21 | 28 | 5 | 13 |
| February | 19 | 26 | 4 | 12 |
| March | 21 | 28 | 5 | 14 |
| April | 19 | 26 | 4 | 12 |
| May | 19 | 25 | 4 | 12 |
| June | 17 | 24 | 2 | 10 |
| July | 16 | 24 | 2 | 10 |
| August | 15 | 22 | 1, 30 | 8 |
| September | 13 | 21 | 29 | 6 |
| October | 13 | 21 | 28 | 5 |
| November | 11 | 20 | 26 | 4 |
| December | 11 | 19 | 26 | 3 |
| | | 2062 | |
| January | 10 | 18 | 24 | 2 |
| February | 9 | 16 | 23 | 1 |
| March | 10 | 17 | 24 | 2 |
| April | 9 | 16 | 23 | 1 |
| May | 8 | 15 | 23 | 1, 31 |
| June | 7 | 13 | 21 | 29 |
| July | 6 | 13 | 21 | 29 |
| August | 4 | 11 | 19 | 27 |
| September | 3 | 10 | 18 | 25 |
| October | 2 | 10 | 18 | 24 |
| November | 1, 30 | 8 | 16 | 23 |
| December | 30 | 8 | 16 | 22 |

| On the Moon you would weigh 1/6th of your weight on Earth. | | | | |
|---|---|---|---|---|
| | | 2063 | | |
| January | 29 | 7 | 14 | 21 |
| February | 28 | 6 | 12 | 19 |
| March | 29 | 7 | 14 | 21 |
| April | 28 | 6 | 12 | 20 |
| May | 27 | 5 | 12 | 20 |
| June | 26 | 3 | 10 | 18 |
| July | 25 | 2 | 10 | 18 |
| August | 23 | 1, 30 | 8 | 17 |
| September | 22 | 29 | 7 | 15 |
| October | 21 | 29 | 7 | 14 |
| November | 19 | 27 | 5 | 13 |
| December | 19 | 27 | 5 | 12 |
| | | 2064 | | |
| January | 18 | 26 | 4 | 10 |
| February | 17 | 25 | 2 | 9 |
| March | 17 | 25 | 3 | 9 |
| April | 16 | 24 | 1, 30 | 8 |
| May | 16 | 23 | 30 | 7 |
| June | 14 | 21 | 28 | 6 |
| July | 14 | 20 | 28 | 6 |
| August | 12 | 19 | 26 | 5 |
| September | 10 | 17 | 25 | 3 |
| October | 10 | 16 | 25 | 3 |
| November | 8 | 15 | 23 | 1 |
| December | 7 | 15 | 23 | 1, 30 |

| *The thin atmosphere does little for protecting the surface.* | | | | |
|---|---|---|---|---|
| | **2065** | | | |
| January | 6 | 14 | 22 | 29 |
| February | 5 | 13 | 20 | 27 |
| March | 6 | 14 | 22 | 28 |
| April | 5 | 13 | 20 | 27 |
| May | 5 | 13 | 19 | 26 |
| June | 3 | 11 | 18 | 25 |
| July | 3 | 10 | 17 | 24 |
| August | 1, 31 | 8 | 15 | 23 |
| September | 29 | 7 | 14 | 22 |
| October | 29 | 6 | 14 | 22 |
| November | 27 | 4 | 12 | 20 |
| December | 27 | 4 | 12 | 20 |
| | **2066** | | | |
| January | 25 | 3 | 11 | 18 |
| February | 24 | 1 | 10 | 17 |
| March | 25 | 3 | 11 | 18 |
| April | 24 | 2 | 10 | 16 |
| May | 23 | 2 | 9 | 16 |
| June | 22 | 1, 30 | 7 | 14 |
| July | 22 | 29 | 7 | 13 |
| August | 20 | 28 | 5 | 12 |
| September | 19 | 26 | 3 | 11 |
| October | 18 | 25 | 3 | 11 |
| November | 17 | 23 | 1 | 9 |
| December | 16 | 23 | 1, 31 | 9 |

| *Just like in space, no sounds can be heard on the Moon.* | | | | |
|---|---|---|---|---|
| **2067** | | | | |
| January | 15 | 21 | 30 | 8 |
| February | 13 | 20 | 28 | 6 |
| March | 15 | 22 | 30 | 8 |
| April | 13 | 21 | 29 | 6 |
| May | 13 | 21 | 28 | 5 |
| June | 11 | 19 | 26 | 4 |
| July | 11 | 19 | 26 | 3 |
| August | 9 | 17 | 24 | 1, 31 |
| September | 8 | 16 | 22 | 29 |
| October | 8 | 15 | 22 | 29 |
| November | 6 | 13 | 20 | 28 |
| December | 6 | 13 | 20 | 28 |
| **2068** | | | | |
| January | 4 | 11 | 19 | 27 |
| February | 3 | 9 | 17 | 25 |
| March | 3 | 10 | 18 | 26 |
| April | 2 | 9 | 17 | 24 |
| May | 1, 30 | 8 | 16 | 24 |
| June | 29 | 7 | 15 | 22 |
| July | 28 | 7 | 14 | 21 |
| August | 27 | 6 | 13 | 19 |
| September | 26 | 4 | 11 | 19 |
| October | 25 | 4 | 10 | 18 |
| November | 24 | 2 | 9 | 16 |
| December | 24 | 1, 31 | 8 | 16 |

| *There is no wind on the moon.* | | | | |
|---|---|---|---|---|
| | 2069 | | | |
| January | 22 | 29 | 7 | 15 |
| February | 21 | 27 | 5 | 14 |
| March | 22 | 29 | 7 | 15 |
| April | 21 | 27 | 6 | 14 |
| May | 20 | 27 | 6 | 13 |
| June | 18 | 26 | 4 | 12 |
| July | 18 | 25 | 4 | 11 |
| August | 16 | 24 | 2 | 9 |
| September | 15 | 23 | 1, 30 | 7 |
| October | 14 | 23 | 29 | 7 |
| November | 13 | 21 | 28 | 5 |
| December | 13 | 21 | 27 | 5 |
| | 2070 | | | |
| January | 12 | 19 | 26 | 4 |
| February | 10 | 17 | 24 | 2 |
| March | 12 | 19 | 26 | 4 |
| April | 10 | 17 | 25 | 3 |
| May | 10 | 16 | 24 | 3 |
| June | 8 | 15 | 23 | 1 |
| July | 7 | 14 | 23 | 1, 30 |
| August | 6 | 13 | 21 | 28 |
| September | 4 | 12 | 20 | 26 |
| October | 4 | 12 | 19 | 26 |
| November | 2 | 10 | 17 | 24 |
| December | 2 | 10 | 17 | 24 |

*Illustration: Klugh, Maria Tales from the Far North*
*(Chicago, IL: A. Flanagan Company, 1909)*

| _The surface area is 14,658,000 square miles or 9.4 billion acres._ | | | | |
|---|---|---|---|---|
| | | 2071 | | |
| January | 1, 31 | 9 | 15 | 22 |
| February | | 7 | 14 | 21 |
| March | 1, 31 | 8 | 15 | 23 |
| April | 29 | 7 | 14 | 22 |
| May | 29 | 6 | 13 | 21 |
| June | 27 | 4 | 12 | 20 |
| July | 26 | 4 | 12 | 20 |
| August | 25 | 2 | 10 | 18 |
| September | 23 | 1, 30 | 9 | 16 |
| October | 22 | 30 | 8 | 16 |
| November | 21 | 29 | 7 | 14 |
| December | 21 | 29 | 6 | 13 |
| | | 2072 | | |
| January | 19 | 28 | 5 | 12 |
| February | 18 | 26 | 3 | 10 |
| March | 19 | 27 | 4 | 11 |
| April | 18 | 25 | 2 | 10 |
| May | 17 | 24 | 2, 31 | 9 |
| June | 16 | 22 | 30 | 8 |
| July | 15 | 22 | 29 | 8 |
| August | 13 | 20 | 28 | 6 |
| September | 12 | 18 | 27 | 5 |
| October | 11 | 18 | 26 | 4 |
| November | 9 | 17 | 25 | 3 |
| December | 9 | 17 | 25 | 2, 31 |

| *The Moon's rotation speed of 10 miles per hour is 100 times slower than the Earth's.* | | | | |
|---|---|---|---|---|
| | | **2073** | | |
| January | 8 | 16 | 23 | 30 |
| February | 6 | 14 | 22 | 28 |
| March | 8 | 16 | 23 | 30 |
| April | 7 | 15 | 21 | 28 |
| May | 6 | 14 | 21 | 28 |
| June | 5 | 12 | 19 | 27 |
| July | 5 | 11 | 18 | 26 |
| August | 3 | 10 | 17 | 25 |
| September | 1 | 8 | 16 | 24 |
| October | 1, 30 | 7 | 15 | 23 |
| November | 28 | 6 | 14 | 22 |
| December | 28 | 6 | 14 | 21 |
| | | **2074** | | |
| January | 26 | 4 | 12 | 20 |
| February | 25 | 3 | 11 | 18 |
| March | 27 | 5 | 13 | 19 |
| April | 25 | 4 | 11 | 18 |
| May | 25 | 4 | 10 | 17 |
| June | 24 | 2 | 9 | 16 |
| July | 23 | 1, 31 | 8 | 15 |
| August | 22 | 29 | 6 | 14 |
| September | 20 | 27 | 5 | 13 |
| October | 20 | 26 | 4 | 12 |
| November | 18 | 25 | 3 | 11 |
| December | 18 | 24 | 3 | 11 |

| *The tallest mountain, Mons Huygens is 18,005 feet (5488 meters) tall* | | | | |
|---|---|---|---|---|
| | | 2075 | | |
| January | 16 | 23 | 2, 31 | 9 |
| February | 14 | 22 | | 8 |
| March | 16 | 24 | 2 | 9 |
| April | 15 | 23 | 1, 30 | 7 |
| May | 14 | 22 | 29 | 7 |
| June | 13 | 21 | 28 | 5 |
| July | 12 | 20 | 27 | 4 |
| August | 11 | 19 | 25 | 3 |
| September | 10 | 17 | 24 | 2 |
| October | 9 | 16 | 23 | 1, 31 |
| November | 8 | 14 | 22 | 30 |
| December | 7 | 14 | 22 | 30 |
| | | 2076 | | |
| January | 6 | 12 | 20 | 28 |
| February | 4 | 11 | 19 | 27 |
| March | 5 | 12 | 20 | 27 |
| April | 3 | 11 | 18 | 25 |
| May | 2 | 10 | 18 | 25 |
| June | 1 | 9 | 16 | 23 |
| July | 1, 30 | 9 | 16 | 22 |
| August | 29 | 7 | 14 | 21 |
| September | 27 | 6 | 12 | 19 |
| October | 27 | 5 | 12 | 19 |
| November | 26 | 3 | 10 | 18 |
| December | 25 | 3 | 10 | 18 |

47

| *Most cultures have used a different sex for their Moon and Sun deities.* | | | |
|---|---|---|---|
| | | 2077 | |
| January | 24 | 1, 30 | 8 | 17 |
| February | 22 | | 7 | 15 |
| March | 24 | 1, 31 | 9 | 17 |
| April | 22 | 29 | 8 | 15 |
| May | 21 | 29 | 7 | 15 |
| June | 20 | 28 | 6 | 13 |
| July | 19 | 27 | 5 | 12 |
| August | 18 | 26 | 4 | 10 |
| September | 16 | 25 | 2 | 9 |
| October | 16 | 24 | 1, 31 | 8 |
| November | 15 | 22 | 29 | 7 |
| December | 15 | 22 | 29 | 7 |
| | | 2078 | |
| January | 13 | 20 | 27 | 5 |
| February | 12 | 18 | 26 | 4 |
| March | 13 | 20 | 28 | 6 |
| April | 12 | 18 | 26 | 5 |
| May | 11 | 18 | 26 | 4 |
| June | 9 | 17 | 25 | 3 |
| July | 9 | 16 | 24 | 2, 31 |
| August | 7 | 15 | 23 | 29 |
| September | 6 | 14 | 21 | 28 |
| October | 5 | 13 | 20 | 27 |
| November | 4 | 12 | 19 | 26 |
| December | 4 | 12 | 18 | 25 |

| _The Moon is not perfectly round, but rather oblong._ | | | | |
|---|---|---|---|---|
| | | | 2079 | |
| January | 2 | 10 | 17 | 24 |
| February | 1 | 8 | 15 | 23 |
| March | 3 | 10 | 17 | 25 |
| April | 1 | 8 | 15 | 24 |
| May | 1, 30 | 7 | 15 | 23 |
| June | 28 | 6 | 14 | 22 |
| July | 28 | 5 | 13 | 21 |
| August | 26 | 4 | 12 | 19 |
| September | 24 | 3 | 10 | 17 |
| October | 24 | 2 | 10 | 17 |
| November | 23 | 1 | 8 | 15 |
| December | 22 | 1, 31 | 8 | 14 |
| | | | 2080 | |
| January | 21 | 29 | 6 | 13 |
| February | 20 | 28 | 5 | 12 |
| March | 21 | 28 | 5 | 13 |
| April | 19 | 26 | 4 | 11 |
| May | 19 | 25 | 3 | 11 |
| June | 17 | 24 | 2 | 10 |
| July | 16 | 23 | 1, 31 | 9 |
| August | 15 | 22 | 30 | 8 |
| September | 13 | 20 | 28 | 6 |
| October | 12 | 20 | 28 | 6 |
| November | 11 | 19 | 26 | 4 |
| December | 11 | 19 | 26 | 3 |

*Everyone is a moon,*
*and has a dark side which he never shows to anybody.*

—Mark Twain

*Illustration: Kim Nguyen*
*strangefamiliarity.tumblr.com*

| *There are periodic moonquakes.* | | | | |
|---|---|---|---|---|
| | **2081** | | | |
| January | 9 | 17 | 24 | 2, 31 |
| February | 8 | 16 | 23 | |
| March | 10 | 18 | 24 | 2, 31 |
| April | 9 | 16 | 23 | 30 |
| May | 8 | 15 | 22 | 30 |
| June | 7 | 13 | 21 | 29 |
| July | 6 | 13 | 20 | 28 |
| August | 4 | 11 | 19 | 27 |
| September | 3 | 9 | 17 | 25 |
| October | 2, 31 | 9 | 17 | 25 |
| November | 30 | 8 | 16 | 23 |
| December | 29 | 7 | 15 | 22 |
| | **2082** | | | |
| January | 28 | 6 | 14 | 21 |
| February | 27 | 5 | 12 | 19 |
| March | 29 | 7 | 14 | 21 |
| April | 27 | 6 | 12 | 19 |
| May | 27 | 5 | 12 | 19 |
| June | 25 | 3 | 10 | 17 |
| July | 25 | 2 | 9 | 17 |
| August | 23 | 1, 30 | 8 | 16 |
| September | 22 | 28 | 6 | 15 |
| October | 21 | 28 | 6 | 14 |
| November | 19 | 26 | 5 | 13 |
| December | 19 | 26 | 5 | 12 |

| *The closest distance to the Sun is 91.3 million miles (147 million kilometers).* | | | |
|---|---|---|---|
| | | 2083 | |
| January | 17 | 25 | 3 | 11 |
| February | 16 | 24 | 2 | 9 |
| March | 18 | 26 | 4 | 10 |
| April | 16 | 24 | 2 | 9 |
| May | 16 | 24 | 1, 31 | 8 |
| June | 15 | 22 | 29 | 6 |
| July | 14 | 22 | 28 | 6 |
| August | 13 | 20 | 27 | 5 |
| September | 11 | 18 | 25 | 3 |
| October | 11 | 17 | 25 | 3 |
| November | 9 | 16 | 24 | 2 |
| December | 8 | 15 | 23 | 2, 31 |
| | | 2084 | |
| January | 7 | 14 | 22 | 30 |
| February | 5 | 13 | 21 | 28 |
| March | 6 | 14 | 21 | 28 |
| April | 4 | 12 | 20 | 27 |
| May | 4 | 12 | 19 | 26 |
| June | 3 | 11 | 18 | 24 |
| July | 2 | 10 | 17 | 24 |
| August | 1, 31 | 9 | 15 | 22 |
| September | 29 | 7 | 14 | 21 |
| October | 29 | 6 | 13 | 21 |
| November | 27 | 4 | 12 | 20 |
| December | 27 | 4 | 11 | 20 |

| *The Moon travels around the Earth at 2,300 miles an hour (3701 kilometers an hour).* | | | |
|---|---|---|---|
| | | 2085 | |
| January | 25 | 2 | 10 | 18 |
| February | 23 | 1 | 9 | 17 |
| March | 25 | 3 | 11 | 18 |
| April | 23 | 1 | 9 | 17 |
| May | 23 | 1, 31 | 9 | 16 |
| June | 21 | 29 | 7 | 14 |
| July | 21 | 29 | 7 | 13 |
| August | 20 | 28 | 5 | 12 |
| September | 18 | 26 | 3 | 10 |
| October | 18 | 25 | 3 | 10 |
| November | 17 | 24 | 1 | 9 |
| December | 16 | 23 | 1, 30 | 8 |
| | | 2086 | |
| January | 15 | 21 | 29 | 7 |
| February | 13 | 20 | 28 | 6 |
| March | 15 | 21 | 29 | 8 |
| April | 13 | 20 | 28 | 6 |
| May | 12 | 20 | 28 | 6 |
| June | 11 | 18 | 26 | 4 |
| July | 10 | 18 | 26 | 3 |
| August | 9 | 17 | 24 | 1, 31 |
| September | 7 | 15 | 22 | 29 |
| October | 7 | 15 | 22 | 29 |
| November | 6 | 13 | 20 | 27 |
| December | 5 | 13 | 20 | 27 |

| *A Harvest Moon is a full Moon that appears during the autumnal equinox.* | | | | |
|:---|:---:|:---:|:---:|:---:|
| **2087** | | | | |
| January | 4 | 11 | 18 | 26 |
| February | 3 | 9 | 17 | 25 |
| March | 4 | 11 | 18 | 27 |
| April | 3 | 9 | 17 | 25 |
| May | 2, 31 | 9 | 17 | 25 |
| June | 30 | 7 | 15 | 23 |
| July | 29 | 7 | 15 | 22 |
| August | 27 | 6 | 13 | 20 |
| September | 26 | 4 | 12 | 19 |
| October | 26 | 4 | 11 | 18 |
| November | 24 | 3 | 10 | 16 |
| December | 24 | 2 | 9 | 16 |
| **2088** | | | | |
| January | 23 | 1, 30 | 8 | 15 |
| February | 22 | 29 | 6 | 14 |
| March | 21 | 29 | 7 | 15 |
| April | 21 | 27 | 5 | 13 |
| May | 20 | 27 | 5 | 13 |
| June | 18 | 25 | 4 | 12 |
| July | 18 | 25 | 3 | 11 |
| August | 16 | 23 | 2, 31 | 9 |
| September | 14 | 22 | 30 | 7 |
| October | 14 | 22 | 29 | 7 |
| November | 12 | 21 | 28 | 5 |
| December | 12 | 20 | 27 | 4 |

| *A Hunters Moon (blood moon) is a full moon that happens after the Harvest Moon.* | | | | |
|---|---|---|---|---|
| | **2089** | | | |
| January | 11 | 19 | 26 | 3 |
| February | 10 | 18 | 24 | 2 |
| March | 12 | 19 | 26 | 3 |
| April | 10 | 17 | 24 | 2 |
| May | 10 | 16 | 24 | 2 |
| June | 8 | 15 | 22 | 1, 30 |
| July | 7 | 14 | 22 | 30 |
| August | 6 | 12 | 20 | 28 |
| September | 4 | 11 | 19 | 27 |
| October | 3 | 11 | 19 | 26 |
| November | 2 | 10 | 17 | 24 |
| December | 1, 31 | 9 | 17 | 24 |
| | **2090** | | | |
| January | 30 | 8 | 15 | 22 |
| February | | 7 | 14 | 21 |
| March | 1, 30 | 9 | 15 | 22 |
| April | 29 | 7 | 14 | 21 |
| May | 29 | 6 | 13 | 21 |
| June | 27 | 4 | 11 | 19 |
| July | 26 | 4 | 11 | 19 |
| August | 25 | 2, 31 | 9 | 18 |
| September | 23 | 30 | 8 | 16 |
| October | 22 | 39 | 8 | 16 |
| November | 21 | 28 | 7 | 14 |
| December | 20 | 28 | 6 | 13 |

*I never really thought about how when I look at the moon, it's the same moon as Shakespeare and Marie Antoinette and George Washington and Cleopatra looked at.*

—Susan Beth Pfeffer, *Life As We Knew It*

*Illustration: The Half Hour Library of Travel,*
*Nature and Science For Young Readers Vol. 3*

| *A full Moon several days before the Spring Equinox causes a Super Full Moon.* | | | | |
|---|---|---|---|---|
| | | | **2091** | |
| January | 19 | 27 | 5 | 12 |
| February | 18 | 26 | 3 | 10 |
| March | 19 | 27 | 5 | 12 |
| April | 18 | 26 | 3 | 10 |
| May | 18 | 25 | 3 | 10 |
| June | 16 | 24 | 1, 30 | 8 |
| July | 16 | 23 | 30 | 8 |
| August | 14 | 21 | 28 | 7 |
| September | 13 | 19 | 27 | 5 |
| October | 12 | 19 | 27 | 5 |
| November | 10 | 17 | 25 | 4 |
| December | 10 | 17 | 25 | 3 |
| | | | **2092** | |
| January | 8 | 16 | 24 | 2, 31 |
| February | 7 | 15 | 22 | 29 |
| March | 7 | 16 | 23 | 30 |
| April | 6 | 14 | 21 | 28 |
| May | 6 | 14 | 21 | 27 |
| June | 4 | 12 | 19 | 26 |
| July | 4 | 12 | 18 | 26 |
| August | 3 | 10 | 17 | 24 |
| September | 1 | 8 | 15 | 23 |
| October | 1, 30 | 8 | 15 | 23 |
| November | 28 | 6 | 13 | 22 |
| December | 28 | 5 | 13 | 21 |

| The craters of the Moon were caused by asteroids and comets. | | | | |
|---|---|---|---|---|
| | | **2093** | | |
| January | 26 | 4 | 12 | 20 |
| February | 25 | 3 | 11 | 18 |
| March | 26 | 4 | 12 | 19 |
| April | 25 | 3 | 11 | 18 |
| May | 24 | 3 | 10 | 17 |
| June | 23 | 2 | 9 | 15 |
| July | 23 | 1, 31 | 8 | 15 |
| August | 21 | 29 | 6 | 13 |
| September | 20 | 27 | 5 | 12 |
| October | 20 | 26 | 4 | 12 |
| November | 18 | 25 | 2 | 10 |
| December | 18 | 24 | 2 | 10 |
| | | **2094** | | |
| January | 16 | 23 | 1, 31 | 9 |
| February | 14 | 21 | | 8 |
| March | 16 | 23 | 2, 31 | 9 |
| April | 14 | 22 | 30 | 8 |
| May | 14 | 22 | 29 | 7 |
| June | 12 | 20 | 28 | 5 |
| July | 12 | 20 | 27 | 4 |
| August | 10 | 18 | 25 | 3 |
| September | 9 | 17 | 24 | 1 |
| October | 9 | 16 | 23 | 1, 30 |
| November | 7 | 15 | 21 | 29 |
| December | 7 | 14 | 21 | 29 |

| *There is a dense inner core.* | | | | |
|---|---|---|---|---|
| | | | 2095 | |
| January | 6 | 12 | 20 | 28 |
| February | 4 | 11 | 18 | 27 |
| March | 6 | 12 | 20 | 28 |
| April | 4 | 11 | 19 | 27 |
| May | 3 | 10 | 19 | 26 |
| June | 2 | 9 | 17 | 24 |
| July | 1, 31 | 9 | 17 | 23 |
| August | 29 | 8 | 15 | 22 |
| September | 28 | 6 | 13 | 20 |
| October | 27 | 6 | 13 | 19 |
| November | 26 | 4 | 11 | 18 |
| December | 26 | 4 | 10 | 18 |
| | | | 2096 | |
| January | 25 | 2 | 9 | 17 |
| February | 23 | 1 | 8 | 16 |
| March | 24 | 1, 30 | 8 | 16 |
| April | 22 | 29 | 7 | 15 |
| May | 21 | 28 | 7 | 15 |
| June | 20 | 27 | 5 | 13 |
| July | 19 | 26 | 5 | 12 |
| August | 17 | 25 | 3 | 10 |
| September | 16 | 24 | 2 | 9 |
| October | 15 | 24 | 1, 31 | 8 |
| November | 14 | 22 | 29 | 6 |
| December | 14 | 22 | 29 | 6 |

| *The outer crust is up to 62 miles (100 kilometers) thick.* | | | | |
|---|---|---|---|---|
| | | 2097 | | |
| January | 13 | 20 | 27 | 5 |
| February | 12 | 19 | 26 | 3 |
| March | 13 | 20 | 27 | 5 |
| April | 12 | 18 | 26 | 4 |
| May | 11 | 18 | 26 | 4 |
| June | 9 | 16 | 25 | 2 |
| July | 9 | 15 | 24 | 2, 31 |
| August | 7 | 14 | 22 | 30 |
| September | 5 | 13 | 21 | 28 |
| October | 5 | 13 | 20 | 27 |
| November | 3 | 11 | 19 | 25 |
| December | 3 | 11 | 18 | 25 |
| | | 2098 | | |
| January | 2 | 10 | 17 | 23 |
| February | 1 | 8 | 15 | 22 |
| March | 2 | 10 | 17 | 24 |
| April | 1 | 8 | 15 | 23 |
| May | 1, 30 | 7 | 15 | 23 |
| June | 28 | 6 | 13 | 21 |
| July | 28 | 5 | 13 | 21 |
| August | 26 | 3 | 11 | 19 |
| September | 24 | 2 | 10 | 18 |
| October | 24 | 1, 31 | 10 | 17 |
| November | 22 | 30 | 8 | 15 |
| December | 22 | 30 | 8 | 15 |

| *A high tide occurs when the Moon is directly overhead, happening twice a day.* | | | | |
|---|---|---|---|---|
| | | | 2099 | |
| January | 21 | 29 | 6 | 13 |
| February | 19 | 27 | 5 | 11 |
| March | 21 | 29 | 6 | 13 |
| April | 20 | 27 | 5 | 12 |
| May | 19 | 27 | 4 | 11 |
| June | 18 | 25 | 2 | 10 |
| July | 17 | 24 | 2, 31 | 10 |
| August | 16 | 22 | 30 | 9 |
| September | 14 | 21 | 29 | 7 |
| October | 13 | 20 | 28 | 7 |
| November | 12 | 19 | 27 | 5 |
| December | 11 | 19 | 27 | 4 |
| | | | 2100 | |
| January | 10 | 18 | 25 | 3 |
| February | 8 | 17 | 24 | 1 |
| March | 10 | 18 | 25 | 2 |
| April | 9 | 17 | 24 | 1, 30 |
| May | 9 | 16 | 23 | 30 |
| June | 7 | 15 | 21 | 29 |
| July | 7 | 14 | 21 | 28 |
| August | 5 | 12 | 19 | 27 |
| September | 4 | 10 | 18 | 26 |
| October | 3 | 10 | 17 | 26 |
| November | 1 | 8 | 16 | 24 |
| December | 1, 30 | 8 | 16 | 24 |

*When we were alive we spent much of our time staring up at the cosmos and wondering what was out there. We were obsessed with the moon and whether we could one day visit it. The day we finally walked on it was celebrated worldwide as perhaps man's greatest achievement. But it was while we were there, gathering rocks from the moon's desolate landscape, that we looked up and caught a glimpse of just how incredible our own planet was.*

—Jon Stewart, *Earth (The Book): A Visitor's Guide to the Human Race*

---

*Illustration: Emmanuel Liais, L'Espace Céleste et la Nature Tropicale, Description Physique de L'univers*

*Sage Liskey is an author, poet, workshop presenter, mental health advocate, and philosopher. He founded the Rad Cat Press in 2010 and is based out of Oregon. The Rad Cat Press is devoted to creating life-changing and accessible publications for the modern world.*

*You can like us on Facebook at www.facebook.com/radcatpress or visit www.sageliskey.com to check out other projects and free downloads also by Sage Liskey and the Rad Cat Press:*

- YOU ARE A GREAT AND POWERFUL WIZARD
- The Happiest Choice: Essential Tools for Everyone's Brain Feelings
- The Happiest Choice: Condensed Edition
- Wine and Poetry Night Year One
- A Sustainability Guide for Everyday Folk
- Community How To
- How To Trim Weed Fast
- The Truthagandist Primer: Effective Information Distribution for Activists
- That Was Zen, This Is Sudoku!
- Surviving the Collapse of Society: Skills to Know and Careers to Pursue

Made in the USA
San Bernardino, CA
09 January 2017